NIGHT CREATURES

Malcolm Penny

Wayland

Dangerous Waters

MONSTERS OF THE DEEP
PIRATES AND TREASURE
VOYAGES OF EXPLORATION
THE WHALERS

Frontiers

JOURNEYS INTO THE UNKNOWN
MAPS AND GLOBES
THE WILD, WILD WEST
THE WORLD'S WILD PLACES

Fearsome Creatures

BIRDS OF PREY
LAND PREDATORS
NIGHT CREATURES
WHEN DINOSAURS RULED THE
EARTH

The Earth's Secrets

FOSSILS AND BONES
THE HIDDEN PAST
THE SEARCH FOR RICHES
VOLCANO, EARTHQUAKE AND
FLOOD

First published in 1995 by
Wayland (Publishers) Limited, 61 Western Road
Hove, East Sussex BN3 1JD, England

Series designer: Jane Hannath
Book designer: David Armitage

Produced by
Roger Coote Publishing
Gissing's Farm, Fressingfield, Eye
Suffolk IP21 5SH, England

British Library Cataloguing in Publication Data

Penny, Malcolm
    Night Creatures. - (Quest! Series)
    I. Title II. Series
    591.51

    ISBN 0 7502 1387 6

Printed and bound in Italy by
G. Canale & C.S.p.A., Turin

Picture acknowledgements
Bruce Coleman 6t/Kim Taylor, 8/Mark Boulton, 12, 16/Kim Taylor & Jane Burton, 33b/Geoff Dore, 36/Jose Luis Gonzalez Grande, 40/M.P.L. Fogden, 43/Hans Reinhard, 44b/Peter A. Hinchliffe; Frank Lane Picture Agency *front cover* r and b/Fritz Polking, 9t/T. Whittaker, 9b/F.W. Lane, 15t/ B. Borrell, 22c/G. Moon, 22b/T. Whittaker, 26b/S. Maslowski, 37b/David Hosking, 41t/E.&D. Hosking; NHPA 13t/Nigel J. Dennis, 19/Stephen Dalton, 24/Agence Nature, 31b/Anthony Bannister, 34/Stephen Dalton, 42/Brian Hawkes; Oxford Scientific Films *front cover* l/Stephen Dalton, 6b/Barrie E. Watts, 7/Partridge Films Ltd, 10/Partridge Films Ltd, 11/Rodger Jackman, 13b/Rafi Ben-Shahar, 14t/Michael Fogden, 18t/Lloyd Nielsen, 21t/Harry Fox, 21b/Tom McHugh, 22t/Breck P. Kent, 23tr/Barrie E. Watts, 23b/Rodger Jackman, 25/R.L. Manuel, 26t/Michael Fogden, 27/Alastair Shay, 28/Wendy Shattil & Bob Rozonski, 29/Stephen Dalton, 30t/Gilbert Grant, 30b/Anthony Bannister, 31t/Konrad Wothe, 33t/Michael Fogden, 35/Rafi Ben-Shahar, 37t/Stephen Dalton, 38b/Jeff Foott,Okapia, 44t/Paul A. Zahl, 45/Norbert Wu, 47/Jeff Lepore; Survival Anglia 38t/Dr F. Köster; Topham Picturepoint 20; Wayland Picture Library 17t/Sarah McKenzie; Zefa 1/Heintges, 4b/Minden, 5t/Aiken, 5b, 17b, 18b, 32/Minden, 39/M. Tuttle. The artwork is by Peter Bull.

CONTENTS

WHY COME OUT AT NIGHT?

Thin skin and huge eyes mark the Madagascar leaf-tailed gecko as a typical night creature.

THE night is an alien world, full of unexplained sounds. Pale moonlight casts deep shadows, distorting familar shapes into things of mystery. The air smells different at night. Away from paths and pavements, humans move with difficulty on uneven ground. A black patch of shadow might be a pool, a hole, or even an unknown creature lying in wait. Into the invisible, we grope and stumble.

Yet around us the darkness is full of life; small, scuttling things whisper and squeak in the leaf litter. A tawny owl shrieks suddenly, an unseen spider's web clingingly caresses our face. Tiny hairs on the back of our neck stand up. Our ears are pricked, but still we cannot hear enough, our eyes open wide, but we can't see. Our nostrils are flared, but our sense of smell can tell us little. We are lost: without the aid of torches and street lights, the night is not our world. How and why do so many other animals make the night their home?

The friendly night

Part of the answer lies in the cool of the night, and part in the darkness.

Right, inset A huge pride of lions gathers to feed on a buffalo in the Maasai Mara, in Tanzania. On the open plains of Africa, where there is little cover, lions hunt mostly at night.

Right The thorn moth flies by night to avoid being preyed upon by birds. But it also has enemies in the darkness - mainly bats.

To a thin-skinned animal that must stay moist, such as a slug or a gecko, the day is a desert, hot and parching. Only at night can they find relief from dry air that would sentence them to death by dehydration. To a moth or a mouse, the cloak of darkness offers safety from daytime predators.

Yet these creatures aren't completely safe: there is a very special set of predators with their own reasons for being about at night-time. They have finely-tuned senses with which to find their prey, even on the darkest night.

Some animals have no alternative to being out at night, because they have to feed all the time just to stay alive. Shrews burn up energy so fast that they can't rest for more than a few minutes at a time without feeding. They hunt all day and all night, looking for worms and beetles, and trying to avoid being caught by hawks in the light and owls in the dark.

A barn owl flies silently on soft wings. It can find its prey in complete darkness by using its specially adapted ears to listen for tiny movements.

During the day, shrews hunt in leaf-litter where they are hidden from view. At night they may come out into the open to find earthworms that come up to the surface.

Eat in peace

For just a few animals, the night is their only chance to feed without fighting, while their rivals are asleep. The douroucouli, or night monkey, lives in the forests of South America, on the eastern slopes of the Andes mountains from Panama to Colombia. It is the only monkey in the world that feeds during the hours of darkness. It sleeps during the day and comes out at night to eat the fruits on the trees. By doing this, the douroucouli can get its share of the freshly ripened fruit before it is all eaten by larger, more aggressive monkeys during the day.

A douroucouli searching for fruit in a forest in Peru, South America.

Hiding from the heat

Not all night creatures are small and defenceless, like shrews. Some are large, and can be dangerous. You might think that because they live in Africa hippos would be used to the sun but, surprisingly,

During the dry season, these hippos in Zambia, Africa, have to crowd together in what little water remains.

they suffer from sunburn. They have to spend the day wallowing in water or thin mud to keep their skin damp and cool. They come out to feed only at night, eating grass on the banks of the pool or river that is their home. There they are safe from the sun, and from predators, too. The only animal big enough to kill hippos is the lion, and lions hunt most often by day.

Hunters in the twilight

A lot of animals are active between day and night, in the dusk and dawn - they are called crepuscular animals. They come out to feed when the hunters of the day are going to sleep, and the night hunters have not yet woken up. Today, the dusk holds new dangers in the form of human hunters, forcing these animals to change their way of life.

Hog deer in India, warthogs in Africa, and peccaries (wild pigs) in South America have all taken to the night, where the risks are great, but not as bad as the threat of a bullet in the dusk.

These warthogs are drinking in a protected area, where they are safe from hunting.

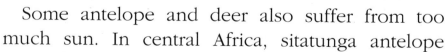

Some antelope and deer also suffer from too much sun. In central Africa, sitatunga antelope have to spend the day hiding in swamps, coming out to feed only as the sun goes down. In South America, a deer called the pudu dies from heat exhaustion within three hours if it cannot find shade.

A sitatunga antelope rests in the shade.

Dinosaur days

Many mammals today are extremely well adapted to life in the dark. Because of this fact, some scientists think that when mammals first appeared on earth, millions of years ago, they might have been night creatures. At that time, the daytime world was ruled by huge reptiles - the dinosaurs, many of which were fierce predators. The only safe time to be out was at night, when most dinosaurs were asleep.

The reptiles needed sunshine in order to keep up their body temperature, and they had to sleep through the cool of the night. But mammals were able to keep their bodies warm day and night. They developed good hearing and a keen sense of smell to help them in the darkness. Their eyes became very sensitive, too. They could not see colours, but there is no colour to be seen at night. To this day most mammals are colour-blind.

The angwantibo of Cameroon, West Africa, is strictly nocturnal. If attacked, it rolls into a ball, covering its head with one arm.

Sun-lovers and night-time feeders

Corals are very unusual creatures, that build hard, cup-shaped skeletons around themselves for protection. Although they are animals, they contain some plant cells which produce food that the coral can use. But these plant cells can only do this in sunlight. The coral animals are soft and easy to eat, so if they emerge from the protection of the hard coral head they will be eaten like grass by passing fish. They have to come out sometimes during the day, but they are always ready to retreat at the slightest sign of danger. In order to feed, they must spread out their tentacles for long periods to collect tiny pieces of food from the water: they do this at night, when most coral-eating fish are asleep.

At night, soft coral tentacles emerge from their stony shelter to collect food.

CAN ANIMALS SEE IN THE DARK?

THE answer depends on what you mean by 'the dark'. Seeing means detecting light, and nothing can see where there is no light at all, such as deep inside a cave. However, many animals can see well when humans can see very little. On what we would consider a very dark night, creatures such as owls, deer, cats, dogs, leopards and antelope can see much more than we can.

Right As the sun goes down, a leopard in Kruger National Park, South Africa, prepares for a night's hunting.

Tarsiers live among the islands of Southeast Asia. Their eyes are so big that they cannot move in their sockets. To overcome this, a tarsier can turn its head almost as far as an owl.

Silent killer

On a moonless night beside the Luangwa River in Zambia, Africa, a female leopard is hunting puku - a type of small antelope. She hunts her prey by scent, stopping often to sniff the breeze. When she catches the scent of a puku herd she changes from silent gliding shadow into a motionless shape - a good imitation of a rock or a tuft of grass. She crouches, completely still, for minutes at a time, before creeping forward on her belly metre by metre. She freezes when one of the antelope raises its head, and then she moves forward again when the nervous herd resumes feeding.

The leopard approaches with the wind in her face, so that the puku will not smell her. The gap between her and her prey closes - from a hundred metres to fifty, twenty, ten... Her movements are so gradual that, to the antelopes' wide eyes, she is just one still shadow among many in the open savanna. Suddenly, she explodes into action. In three giant strides, she covers the last 8 metres in a split second. Her chosen victim knows nothing until the leopard's great weight drags it to the ground by the neck. In seconds, the puku is dead, throttled by the hunter's powerful jaws.

The rest of the herd scatters, darting between trees and bushes. Meanwhile, prowling hyenas listen, sniff the air, and watch for any chance of a meal. They will steal the leopard's kill if they can, but they are also fierce hunters - and they, too, can see perfectly at night.

A leopard usually disembowels its prey, to make it lighter before carrying it up a tree, out of reach of hyenas.

The wide-open eyes of a spotted eagle owl in Uganda, Africa.

Specialist eyes

Nocturnal animals have such good night vision because of the structure of their eyes. The main parts of an eye are the lens and the retina. The lens lets light into the eye, and the retina turns the light into signals that are sent to the brain. The bigger the lens, the more light it can let in, and the more sensitive the retina, the better the signal it can send to the brain. Animals that need to see in dim light must have a wide-open lens, and a very sensitive retina.

Rods and cones

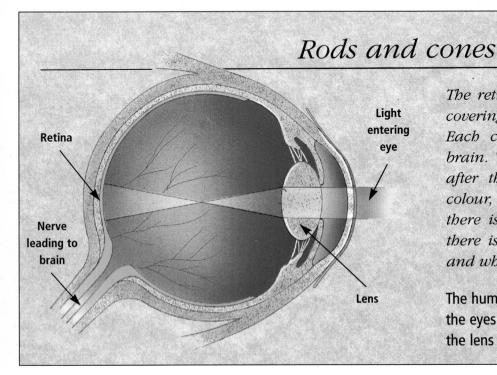

Retina

Nerve leading to brain

Light entering eye

Lens

The retina is a sheet of light-sensitive cells covering the back of the inside of the eyeball. Each cell is connected by nerves to the brain. There are two kinds of cell, named after their shape. Cone cells can detect colour, but rod cells detect only whether there is any light or not, and how much there is: in other words, they see in black and white and shades of grey.

The human eye works in much the same way as the eyes of other mammals. Light is focused by the lens on to the sensitive retina.

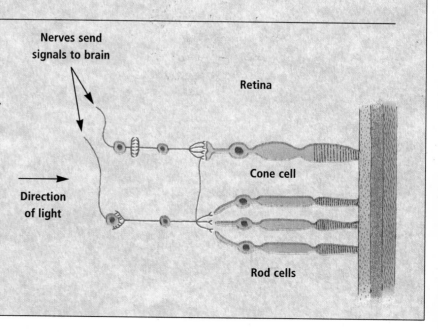

Apes, cats and colour vision

Most mammals are colour-blind. Only the higher primates - that is, monkeys, apes and humans - can see in full colour. The only exception is the night monkey, which is nocturnal, so colour vision would be no use to it. Its eyes are like those of the prim-itive mammals, with no cone cells in their retinas.

An unexpected result came from an experiment with cats, in which scientists measured the signals from their eyes to their brain. It turned out that a cat's eye contains cones, and can respond to coloured light. However, when the cats were test-ed to see whether they could learn to choose between different colours, they failed on every test. It seems as though cats' brains aren't able to make use of all the information that their eyes give them.

In bright light, the pupils of the eye close down to prevent the retina from being damaged by too much light. A cat's pupils become narrow vertical slits.

Each cone cell has its own connection to the brain, but several rods share a nerve. This means that colour vision produces a very clear picture, though it needs plenty of light, while black-and-white vision works with little light, but produces a picture that is not quite so clear. Night creatures have very few cones in their retinas, or often none at all, so that they can see in dim light, but get a rather blurred picture.

The arrangement of rods and cones in a section of the retina of a human, monkey or ape.

Nerves send signals to brain

Retina

Direction of light

Cone cell

Rod cells

A built-in mirror

f you see a dog, a deer, or a cat in the headlights of a car, its eyes shine back at you with a red or green glow. This 'eyeshine' is produced by a layer of reflecting cells, like a mirror at the back of the eye, behind the retina. The mirror is called the tapetum.

The tapetum recycles the light entering the eye, by making it hit the retina twice, so that each rod receives two signals. This blurs the picture but it also makes it brighter.

Big eyes and bendy necks

To an owl, a woodland at night is far from dark. It can see the trees and the ground - especially lighter patches in clearings between the trees. Its view from its perch 3 or 4 m above the ground is quite blurred, and it certainly wouldn't be able to see a shrew or a mouse. Instead, it listens, turning its head from side to side, silently identifying the many rustles and squeaks from the forest floor.

When it recognizes the rapid, bustling movements of a shrew hunting among the dry leaves, the owl turns it head towards the sound, and becomes very still. Silently, it glides on its soft wings, down towards the sound. As it flies closer to its prey, the owl's eyes come into focus. It sees the shrew clearly now, and with a last-second adjustment it turns its talons so that they catch the shrew from the side, giving it no chance of escape.

A listener might hear the crash of the owl's impact, and the shrew's final squeak - and then silence, broken only by the faint crunching and tearing sounds of the owl eating its prey.

Geckos spend most of the day hiding, motionless, in quiet, shady places. They still need to be able to see to avoid predators, and their eyes are specially designed for this. The gecko's pupil has notches in it, so that even when it is closed it lets in a little light - enough to detect movement.

In daylight, a barn owl's eyes are almost completely closed to shut out most of the light. Even so, it is still too dazzled to see properly.

The last of the light

Nightjars feed by sitting on the ground and watching the night sky. They take off to catch flying insects when they see them silhouetted against the sky. They can do this even on the darkest of nights. This is because there is always some light at night, such as the faint glimmer of the stars, even when the sky is covered in clouds. The eyes of nightjars can respond to these very faint levels of light.

A spotted nightjar in Queensland, Australia, watches the sky.

Owls are well-known for 'seeing in the dark', although their eyes are quite short-sighted. To an owl, for example, starlight is as bright as full moonlight is to us. The tawny owl is completely nocturnal, and its eyes are ten times as sensitive to light as ours. In order to gather as much light as possible, an owl's eyes are as big as the space inside its skull will allow. They are also a very unusual shape. The lens at the front and the retina at the back are very large, but the space in between is pinched in by the bones of the skull, so the owl's eye is shaped like a tube.

Because of this, an owl cannot move its eyes in their sockets like other animals, to look from side to side. Instead, it has a very flexible neck, so that it can swivel its head through about 270 degrees. This means that owls can look almost directly behind them.

Using their flexible necks, two barn owls take a good look at the photographer. Moving their heads sideways helps them to judge distance.

Watching animals at night

Although natural sunlight is sometimes called white light, it is actually made up of all the colours of the rainbow. Our eyes can see all of these colours, but nocturnal animals see mostly green light, and hardly respond to red or blue. So, the best way to watch them is to use a torch covered with a red or blue filter.

However, night creatures have other highly developed senses to help keep them safe in the dark. *You will need to be very quiet, and to keep the breeze in your face so that it does not carry your scent towards the animals you want to watch.*

A leap in the dark - a bush baby in full flight. Bush babies jump among the trees at night, guided by their huge eyes. Their bushy tails help them to balance as they leap from branch to branch.

ON THE SCENT

ALMOST everything has a smell, and smells are closely linked with memory. In the dark, when seeing is difficult, and sounds may be drowned out by the rustle of leaves or falling rain, scents can give vital information. Even more important is remembering what the scents mean.

Some scents do not have to be remembered because they directly affect an animal's behaviour. They are called pheromones and they are found throughout the animal kingdom, up to and including humans. The best examples are found among insects, where tiny amounts of pheromones can affect an animal's actions dramatically.

Follow the scent

A female vapourer moth cannot fly. She sits on the trunk of a tree or another convenient perch, and pumps out a pheromone into the night. A male vapourer can detect the scent from as far as 8 kilometres away.

A timberman beetle, found in Britain on dead pine trees and in stacks of pine logs. Males can have antennae five times as long as their body. They use them to detect the scent of females.

A male vapourer moth has used his comb-like antennae to detect the scent of a pale, wingless female.

As soon as he picks up even the faintest scent, he begins to fly towards it. He has no choice - the pheromone has affected his behaviour. He follows the scent gradient, always flying towards the place where the scent is strongest, until he finds the female and mates with her.

Dung beetles, too, fly at night along a scent gradient, but only to find food. They can detect the smell of fresh dung from many kilometres downwind. Once they have fed, and gathered some stores for their nest, the scent has no further effect on them, unlike the male vapourer moth, who will fly on to find another mate.

Can birds smell?

There are only three birds that have a sense of smell - kiwis, kakapos and oilbirds. They are all night creatures.

The kiwi, which lives in New Zealand, is unable to fly. It feeds in leaf litter on the ground, sniffing out insects and earth-worms with its long, flexible bill. Unlike all other birds, its nostril openings are right at the tip of its bill.

A kiwi sniffs out food in the darkness.

Snakes smell in stereo

Snakes and lizards test the air around them by collecting tiny samples of scent on the tips of their forked tongues. When they pull their tongue in, its tips fit into two depressions inside the mouth, called Jacobson's organ, after the man who discovered them. The pits are lined with scent-detectors, and because there are two of them, the reptile can smell in stereo, to build up a three-dimensional picture of its surroundings.

A coachwhip snake from the western United States.

African wild dogs hunt mainly at dawn and dusk, but when they go out at night scent is their main means of finding prey.

Kiwis have very small eyes, but quite large ear openings, suggesting that they can hear quite well. But their main means of finding food is by scent.

The kakapo is another flightless bird from New Zealand. It is sometimes called the 'owl-parrot' because, although it is a ground-living parrot, the radiating feathers round its eyes make it look rather like an owl. It feeds on grass and mosses, and also fruits and flowers, which it finds by their smell. It also sometimes eats lizards, but probably only when it steps on them by chance.

A kakapo is well-camouflaged among the mosses of the scrub-forest where it lives.

The oilbird is the world's only nocturnal fruit-eating bird. It uses its sense of smell to search out food, such as oil nuts and the scented fruits of laurel. Oilbirds pick them on the wing and then return to their caves to digest the food during the day.

WHISKERS, WEBS AND FINGERS

A water scorpion catches its prey in one or both of its front legs, which snap shut instantly when anything touches their sensitive bristles.

THE sense of touch is very important to night creatures. The tips of their limbs are often covered by claws or hooves, which are not very sensitive. But their whiskers and other hairs can detect the faintest touch.

Hair-triggers

People often say that an animal's whiskers enable it to judge the width of openings through which it can squeeze, but this is not their main function. Every individual hair has a nerve at its base which is sensitive to movement: if a whisker is touched, a message flashes to the brain.

The splendid whiskers of this house mouse enable it to detect obstacles close to it as it moves about at night.

The whiskers around the mouth of a predator - whether it is a mountain lion or a mongoose - act as triggers, so that the animal can turn and snap its jaws as its prey makes a last desperate effort to escape. Even mice and shrews, not usually thought of as bloodthirsty predators, use their whiskers to sense the movement of insects as they forage during the night.

Some species of crabs have lightning reactions that help them to trap their prey. Crabs living in deep water, or in caves, sense the movements of small crustaceans as they feel for them in the dark. The inner surface of each claw is covered with sensitive bristles that make it snap shut at the lightest touch.

This mysterious deep-sea crab hunts by touch in total darkness, using tiny bristles on its claws. The spines that cover its body probably put off other animals that might try to eat it.

A green snake-locks anemone lies in wait for prey in a shallow rocky gully.

Twangs and thuds

Most spiders that use webs build them at night, and many of their victims are night-flying insects. The spider lurks close by, but it is not watching the web - it is feeling it. With its forefeet resting on the strands of silk, it waits for the tell-tale vibration of an insect hitting the web, and the thrumming in the strands as its prey struggles to escape. From the strength of the signal and its effect on the two strands it is touching, the spider can tell how far away the insect is and whereabouts in the web.

Unable to use any other senses, the spider catches its prey by touch.

Snakes also use touch, but in their case it replaces hearing. All snakes are deaf, but they are alert for the vibrations in the ground that warn them of the approach of a large creature.

Right An orb web spider in the Gambia, West Africa, weaves a zig-zag of white silk into its web. This makes it easy for birds to see, so that they will not fly into it and wreck the spider's work.

A striped racer snake from Costa Rica 'listens' for approaching footsteps by pressing its body to the ground.

A party of raccoons sorts through the refuse in a back-yard in Florida, in the United States.

Fingers and false feelers

The raccoon, which lives in America, from southern Canada to Panama, is the best example of a nocturnal animal that feels for its food with sensitive fingers. Although they have now become experts at raiding dustbins and stealing from chicken runs, raccoons in the wild feed in shallow muddy water, groping about until they find a crab or crayfish to eat. Sometimes, though, they touch the trigger hairs on a crab's claw - the raccoon gets a nasty nip and the crab has the last laugh.

Moths versus spiders

A spider's sticky web is ideal for catching moths. When the spider feels a moth hitting the web, it scuttles out to tie the moth up with silk before sucking out its body fluids. But moths have a defence against this sticky end: the scales on their wings are very loosely attached, so they come off easily and leave the moth free to fly away.

Some spiders, such as the large, tropical orb-web spiders, have a secret weapon to counter the moths' defences. They build an extra piece of web that stretches down like a ladder below the main web. When the moth tries to fall free, it runs out of loose scales before it reaches the bottom of the web, and the spider gets its meal.

USING SOUND IN THE DARKNESS

A good sense of hearing is vital to nocturnal animals, either to find their own prey or to avoid becoming something else's.

Hunters and hunted in stereo

To hear in stereo you need two ears. Each ear receives a slightly different signal, so that the brain can tell whether the sound is coming from the left or the right, in front or behind. Because both ears are usually the same height from the ground, it is hard to guess how high up the source of a sound is. Owls have solved this problem by having one ear higher in their head than the other.

A greater horseshoe bat chasing a moth. The bat can't see its prey, but it knows where the moth is by listening for the echos of its own high-pitched squeaks bouncing off the moth's body.

A cottontail rabbit in Colorado, USA, listens for danger in the early-morning sun.

Bad weather for rabbits

Rabbits usually feed from dusk to dawn, and they rely on their sense of hearing to warn them of danger. If it is raining or very windy, the noise of the weather can mask the sounds of an approaching predator, such as a fox, so the rabbits stay underground and come out to feed in the morning. Country people sometimes say that rabbits feeding in daylight means that rain is on the way. In fact, it is a sign that it was raining or windy in the night, forcing the rabbits to stay in their burrows.

Big ears

In the dry country of the south-western United States, the kit fox and the Californian leaf-nosed bat are two nocturnal creatures that are real specialists at finding their prey by listening.

Kit foxes are delicately built, with slim legs and very big ears. They feed on large insects and small rodents, such as kangaroo rats. They hear their prey from a long way away, then use their noses to find exactly where it is, and their whiskers to guide the final snap of their jaws as they catch it.

The leaf-nosed bat also has enormous ears, which point down towards the ground as it flies silently through the dark. They are so sensitive that the bat can hear the footsteps of crickets on the desert floor. Only when it has heard one does it switch to echo-location for the final attack.

Left A Californian leaf-nosed bat roosting during the day. Its enormous ears can hear an insect's footsteps on the desert floor.

In the Kalahari National Park, South Africa, a honey badger listens for the sounds of grubs moving underground.

The scarab and the honey badger

The honey badger, or ratel, is a nocturnal animal that lives in Africa south of the Sahara. One of its favourite foods is honey, as its name suggests. But it is also partial to the grubs of a large dung beetle, or scarab, which feeds on elephant droppings.

A young aye-aye (a type of lemur) in Madagascar. It listens for the movements of grubs under tree-bark, then gnaws into the wood and winkles out the grubs with its long, thin middle fingers.

Messages in the dark

The sounds that animals hear in the night are not always those of prey or predators. Many come from their own species. They may be contact calls, as one animal tries to find others of its kind; territorial calls, warning other animals to keep their distance; or mating calls at breeding time.

The beetle lays its eggs in balls of dung in a chamber nearly a metre underground. Each grub develops inside a hard ball of earth the size of a tennis ball. The ratel can hear the grubs moving inside the balls, and digs them up with its powerful claws.

African scarabs rolling a ball of dung to their underground nest. They will lay an egg inside the ball and coat it with mud.

A red howler monkey calls to others of its kind in the South American jungle.

The South American night is often disturbed by the calls of tribes of howler monkeys. It is important for the tribes to be well spaced out, so that each has enough space to find food without having to waste time fighting with its neighbours. If the calls come from too close, the tribes move apart.

In Africa, the howling and chattering of feeding hyenas is one of the most blood-curdling sounds of the night. Its purpose is different from the calls of the howler monkeys: rather than keeping them away, it invites other hyenas to share the feast.

How quiet is the night?

Night-time is much quieter than the hustle and bustle of the day, but to night creatures it is full of coded signals. Even small sounds may have great significance. The clack of an owl's bill in a tropical forest might sound like a snapping twig, but to another owl it is full of information - it may be calling for a mate or proclaiming its ownership of a territory.

The calls of other animals, such as the growl of a puma or the roar of howler monkeys, do not interfere with the messages between owls.

Crickets and tree frogs also call at night, when they are looking for mates. In the forests of Costa Rica, the continuous noise of thousands of frogs and millions of crickets would seem to make any communication impossible, but in fact each species can hear only the calls of its own kind.

When is it better to be silent?

Male crickets in the grasslands of Texas, in the United States, call to attract mates. Females crawl up to the loudest voice they can hear of their own species. The males spread out, so that their call will not be confused with that of a neighbour. If another male comes too close, there will be a fierce fight until one of them moves away. Unfortunately for the male crickets, another animal can also hear their calls: it is a parasitic fly, which attaches a larva to the crickets' skin. The larva burrows into a cricket and eats it from the inside out. When a male cricket calls for a mate, it is also inviting an early death.

A brown-and-gold tree frog in Costa Rica, South America, sings to attract a female. The air sac under its chin makes the calls louder.

Robbing the hunters

In 1987, 4,000 hippos died in Zambia's Luangwa National Park from an outbreak of the disease anthrax. Hyenas ate the hippo meat and used it to feed their cubs, so that more cubs survived than usual. This was nearly a disaster for the leopards in the park. The hyenas were everywhere, every night, calling their packs together to rob leopards of their prey before they had time to take it to safety up a tree. Now the leopards have learned to catch smaller prey, so that they can escape with it more quickly, but the hyenas still rob many of them.

A party of spotted hyenas feeding on a giraffe. They may have killed it themselves or robbed other hunters.

Some males have a way of avoiding this danger. They come to the breeding grounds, but they do not call. Instead, they wait near another male with a loud voice. When a female comes near, the silent male mates with her himself. Although silent males do not mate as often as singers, they keep out of fights and avoid being infested with parasites. This means that they live longer, and eventually have the same chance of breeding.

Aerobatics

On the darkest night, bats can fly around and avoid obstacles such as leaves, branches and power lines. While doing this, they can also catch flying insects to eat. Scientists have used special cameras to film bats flying in the dark. In one test, a horseshoe bat with a 40-cm wingspan flew through a net in which the holes were only 14 cm square. The net was made of transparent nylon threads, 0.08 mm thick. Amazingly, the bat folded back its wings and flew through the net without touching the threads.

A fishing bat turns its claws forward, ready to snatch up a small fish it has detected near the water surface. If a fish's fin sticks up as little as half a millimetre above the surface, the bat can find it by echolocation.

People sometimes use the expression 'as blind as a bat'. In fact, bats aren't totally blind; they are able to tell the difference between light and darkness. But that is about all they can see. It isn't the bats' eyes that guide them as they flap and swoop through the darkness: they are using their voices and their ears.

Echoes in the dark

Bats find their way around by echo-location - they make very short, high-pitched calls, and listen for the echo from anything that might be in front of them. Bats make their calls through their noses, which often have complicated flaps and bulges to direct the sound, and they listen for the echoes with large, specially adapted ears.

So that they don't deafen themselves with their own loud calls, bats can switch off their ears just for the very short time it takes to make each call. When scientists recorded and measured the calls of a hunting bat, they found that each was no more than 15 thousandths of a second long.

The greater horseshoe bat get its name from the shape - called a leaf - on its nose. Bats use the leaf to focus the sounds they produce.

The call of the vampire

Vampire bats make very quiet calls to find their way around, before creeping close to their victim and licking its blood from a tiny cut made by their sharp teeth. Cattle and birds are regularly bitten, but dogs and cats have such good ears that they can hear the bats coming, and get out of the way.

A vampire bat laps blood from the snout of a sleeping pig. The pig did not feel the bite because the bat's teeth are so sharp.

Dark caves and deep waters

Other animals use a similar system to find their way about. Birds that live in dark caves, such as swiftlets and oilbirds, and mammals that live in deep water, such as dolphins and killer whales, use echo-location to find their nests or their prey.

A cave swiftlet on its nest in a cavern in La Digue, Seychelles. The nest is made from fine plant material stuck together with the bird's saliva.

An oilbird prepares to leap into the darkness from its nesting ledge. As soon as it is in the air it will start to make echo-location calls.

Cave swiftlets are like small swallows, living in the tropics. They hunt insects by sight during the day, but they roost and nest deep inside caves, beyond the reach of daylight. As they fly from the light into the dark, their calls change, from a twittering sound to a harsh buzzing and clicking noise. The echoes of these sounds help the swiftlets to find their way about.

The oilbirds of Central America and the Caribbean roost in caves by day, and go out at night to find food. They have sharp, hooked bills, like birds of prey, and they use them to collect fruit, plucking oil-nuts from palms as they fly by. Oilbirds have large eyes, and they fly by sight. But when they return to their caves they begin making buzzes and clicks, like cave swiftlets, to find their way back to their roosting place in the darkness.

Listening to dinner

Killer whales are experts at echo-location. They send out loud clicks from the front of their head near their nose, and receive the echoes in the point of their jaw, where channels full of oil carry the sound to their ears. Some of the sounds killer whales make are a type of language which the whales use to communicate with each other. Some other sounds are like the clicks and buzzes of cave-living birds.

A young killer whale will learn from its mother the special language used only by their family, or pod. All killer whales use the same echo-location sounds, but each pod has its own dialect, which is different from all other dialects.

38

Flying by sight

Not all bats use echo-location to fly and hunt. There are some that fly by sight. They are the fruit bats, sometimes called flying foxes because of their fox-like faces and reddish fur. They live in the tropics and subtropics in parts of Europe and Asia. Some of them are very large; the biggest, from Southeast Asia, has a wingspan of 1.5 m.

At sunset, fruit bats go out looking for trees with fruit, or flowers with plenty of nectar. They have large eyes, and the surface of the retina at the back of each eye has bumps on it to increase its area. This enables it to collect more light and help the bats see more clearly. Even so, fruit bats do have problems: on dark evenings they often crash into electricity cables because they can't see well enough to avoid them.

An epauletted African fruit bat. Fruit bats, or flying foxes, do not need the elaborate ears and noses of the bats that use echo-location.

A killer whale's echo-location system is so accurate that it can tell a cod from a salmon at a range of more than 20 metres. There may be two explanations for this amazing ability. Either the whale is able to judge the shape of the fish, or it can sense the different way in which the two fishes move. Whatever the reason, if they have the choice, killer whales will always choose salmon, which they like much better than cod.

ARE THERE MORE THAN FIVE SENSES?

THE lives of some animals are very different from our own, and the senses they possess can be hard for us to understand. Perhaps the strangest is the world of the pit vipers, which can detect and hunt their prey in total darkness, even when they are underground.

Heat-seeking snakes

A pit viper has two heat-sensitive organs in its face, located in pits just below the eyes. These organs are sensitive to infra-red radiation, which is like light that is too dim for us to see, or heat so faint that we can't feel it. No matter how well-insulated an animal is, its nose and eyes, and the lower part of its belly, will give off heat. These are the parts which the snake can detect.

When a small animal dies, it cools down very quickly and stops giving out heat. If a pit viper's prey could run any distance after being bitten, the snake might not be able to find it again in the darkness. To prevent this from happening, the venom of pit vipers is among the strongest of all snakes. When they bite, their prey drops dead almost instantly.

A hog-nosed viper in Costa Rica swallows a frog whole. Like most other snakes, it can dislocate its lower jaw to make room for large meals to enter its stomach.

Above An albino diamond-backed rattlesnake from Arizona, USA, showing the heat-sensitive pits below its eyes.

RightThe image of a mouse a pit viper 'sees' using its pit organs could be something like this. The warmest parts - the brain and lungs - are shown as yellow or red.

A compass in the head

Many animals have a magnetic sense. Some sharks, especially those that feed on the sea bottom, can detect tiny changes in the earth's magnetic field caused by prey animals buried under several centimetres of sand.

The magnetic field of the earth is what makes a compass needle point to the north. People have wondered for years how migrating birds find their way across oceans and continents, especially at night. They were sure that birds could navigate by observing the position of the stars, but they could not explain how the birds managed to find their way on cloudy nights. When some scientists suggested that birds follow the earth's magnetic field, people laughed. Recently, it has been proved that birds that migrate long distances really are sensitive to the magnetic field and they use it to find their way.

Left House martins gather on telegraph wires as they prepare to migrate from Europe to Africa. Their brains contain tiny magnetic crystals with which they can detect the earth's magnetic field. Recently, similar crystals have been found in human brains, so perhaps we too have a compass in our heads, like migratory birds.

Electric fish

In the murky depths of some rivers in Africa and South America, there are electric fish called mormyrids, which have a very strange way of finding prey. They send out electrical pulses, up to 300 per second, and detect the way in which those pulses are changed by their surroundings - including their prey. The nearest thing in human experience is a metal detector, which works in much the same way.

Some electric fish do even more. Electric eels, some rays and electric catfish can send out a shock that is powerful enough to stun or even kill other fish. Some eels, known as gymnotids, can detect tiny electrical impulses produced by the movement of the muscles of their prey. In total darkness, they create a picture of their world using a sense that we can barely imagine.

Some of the groups of muscles in an electric eel's body have become modified to act as batteries, producing a powerful electric shock with which it can stun or kill its prey.

Lighting up the night

Glow-worms and fireflies can create their own light, to signal to each other at night, usually to attract a mate. Different species produce light flashes of different lengths, so that they can tell each other apart.

In Florida, in the United States, some fireflies cheat in order to trap their neighbours. One species sends out flashes that mimic the signal sent by females of another species. When eager males fly down to mate with what they think is a female of their own species, the cheating firefly catches and eats them.

Above This Jamaican firefly has been tipped on to its back to show the three segments of its abdomen where it produces light.

Right Like fireflies, European glow-worms are really beetles. The light is made by a chemical reaction.

Deep-sea lanterns

Deep down in the sea, below about 100 m, there is almost no light at all because the sun's rays can't reach that far through the water. However, some of the fish that live there can produce their own light.

Hatchet fish, for example, make light to protect themselves from predators, which might be able to see them silhouetted against the faint light from the distant surface. They have glowing spots on their underside to act as camouflage, breaking up their outline so that their shape is harder to identify from below.

The lantern fish uses light as bait. It has a long filament that sprouts from its head and dangles in front of its mouth. At the end of the filament is a knob that glows with its home-made light. When small fish come to investigate the mysterious light in the darkness, they are quickly swallowed up in the cavernous mouth of the lantern fish.

In the darkness of the deep ocean, a hatchet fish gazes upward with huge eyes. Glowing plates on its underside are intended to confuse predators hunting from below.

GLOSSARY

Coral A small animal related to the sea anemones, that builds stony colonies to live in. A lot of colonies together is called a coral reef.

Crepuscular A word describing animals that are active in the twilight period between light and darkness.

Dehydration Losing water.

Diurnal Active during the day.

Echo-location Sensing the surroundings by making sounds and listening for the echo from objects.

Forage To look for food.

Insulated Protected from losing or gaining heat.

Iris The coloured ring in the front of the eye.

Larva The young form of an insect, after it has hatched from the egg.

Leaf litter The dead leaves and twigs that cover the floor of a wood or forest.

Lemur A form of primate, found mostly in Madagascar.

Lens The transparent part of the eye that focuses light on to the retina.

Mammal An animal that feeds its young with milk.

Nocturnal Active at night.

Parasitic Feeding on the body or blood of a living animal.

Predator An animal that kills and eats other animals.

Prey The food of a predator.

Pupil The hole in the middle of the iris, through which light passes to the lens.

Reptile An animal that is covered in scales, and cannot control its own body temperature from within.

Retina The sheet of cells inside the eye that is sensitive to light.

Rodent An animal like a rat or a mouse, with sharp front teeth.

Roost To sleep. The word is mainly used in the case of birds.

Subtropics The area between the tropics and cooler parts of the world.

Tentacles Flexible arms with which some animals catch their prey.

Tropics The hottest parts of the world. They are on either side of the equator, extending north as far as the Tropic of Cancer and south to the Tropic of Capricorn.

Wallowing Lying in mud or water to keep cool.

FURTHER INFORMATION

BOOKS

The Barn Owl by Mike Read and Jake Allsop (Blandford, 1994)

Bats by Phil Richardson (Whittet Books, 1994)

Birds of the Night by Jean de Sart (Belitha Press, 1994)

Coral Reef by Barbara Taylor (Dorling Kindersley, 1992)

Life in the Dark by Joyce Pope (Heinemann, 1991)

Mammal by Steve Parker (Dorling Kindersley Eyewitness Guides, 1989)

Night Creatures (Time-Life Books, 1986)

Night Life; the Secret World of Nocturnal Creatures by Malcolm Penny (Boxtree, 1993)

Spiders of the World by Rod and Ken Preston-Mafham (Blandford Press, 1984)

A number of titles in the Dorling Kindersley 'Eyewitness Guides' series also contain some relevant information about the creatures described in this book. They include ***Spider***, ***Butterfly and Moth***, ***Amphibian***, ***Reptile***, ***Insect***, ***Fish***, ***Cat*** and ***Dog***.

Coyotes live in almost all of the USA. They howl to signal their presence to others of their kind, and their haunting calls can be heard for several kilometres, especially at night.

INDEX